# MONTREAL'S EXPO 67

BILL COTTER

ARCADIA
PUBLISHING

Published by Arcadia Publishing
Charleston, South Carolina

Printed in the United States of America

Library of Congress Control Number: 2015953564

For all general information, please contact Arcadia Publishing:
Telephone 843-853-2070
Fax 843-853-0044
E-mail sales@arcadiapublishing.com
For customer service and orders:
Toll-Free 1-888-313-2665

Visit us on the Internet at www.arcadiapublishing.com

*To Carol: Thanks, once again, for your help and support.*
*If I ever do get a time machine, we're going back to*
*Expo 67 so you can see why I'm so excited about it.*

# CONTENTS

# Acknowledgments

This is the 10th world's fair book I have written for Arcadia. I have enjoyed doing each of them, but working on this book has been a special treat for me. I am lucky to have been able to go to a number of world's fairs over the years and have enjoyed each of these international expositions. Expo 67 was unique, though.

Hosting a world's fair is not a trivial matter, for building these events inevitably introduces a number of problems. In most cases, large sections of the city have to be cleared to make way for the pavilions, usually displacing some rather unhappy residents and businesses. Traffic woes can last for years, both during the construction process and while the event is open. Prices can soar at local restaurants; parking becomes impossible; and when the fair finally ends, it has almost always lost money, leaving the city in debt for years to come.

It is not surprising, then, that the people living close to an expo are often not too happy with their temporary neighbor and avoid it as much as possible. In extreme cases, such as the opening of Expo 2015, they have actually rioted in the streets in protest of the event. This negative reaction stands in stark contrast to what I experienced during Expo 67.

If ever a city loved and embraced its fair, it was Montreal showing its love for Expo 67. Yes, there were political battles over the location, the financing, and other issues, but by the time the gates opened the problems were forgotten and the fun began. Expo 67 was a wonderful event, celebrated by people across the country and from around the world, but especially by the people of Montreal. It was a joy to be there then and to tell its story now.

My thanks to those who helped with their reviews of the manuscript and suggestions: Helen Kirzeder Schulte, Gerard Schulte, Katie Carver, David Carver, Joseph Aspler, James S. Flanagan, Randy Lopes, Clément Drolet, Louis Turgeon, Mark Strobl, and especially my wife, Carol, who has supported me every step along the way. A special thanks to Roger La Roche for all of his work on the French version—I couldn't have done it without him.

All photographs are from the author's collection unless otherwise noted.

—Bill Cotter
worldsfairphotos.com
August 2016

# INTRODUCTION

*"Être homme . . . c'est sentir, en posant sa pierre, que l'on contribue à bâtir le monde."*
*(To be a man . . . is to feel that, through one's own contribution one helps to build the world.)*

—Antoine de Saint-Exupéry

What came to be known as Expo 67 began in 1958 when Canadian politicians began exploring ways to celebrate the 100th anniversary of the Federation's founding in 1867. The first public announcement that they were considering a world's fair was made by Sen. Mark Drouin on October 28, 1958, at Expo 58 in Brussels, but the planners had to overcome some considerable roadblocks before their dreams could become a reality.

The first challenge was obtaining approval from the Bureau International des Expositions (BIE), a Paris-based organization that had been formed in 1928 to regulate world's fairs. Several other nations were also interested in hosting a fair in 1967, the target year, and when the BIE announced its decision in 1960, Canada had narrowly lost to the USSR. The Canadian delegation began working on other, less grand, plans for the centenary, but to everyone's amazement and joy, the USSR decided in 1962 that it would not hold its fair, citing concerns about security and cost. Although other countries immediately expressed interest in hosting a 1967 fair, on November 13, 1962, the BIE announced that Canada would be the host of "the Universal and International Exhibition of 1967."

A new federal agency, the Canadian Corporation for the 1967 World Exhibition, was formed to design and operate the fair. One of the first obstacles it had to overcome was where to put it. Montreal mayor Jean Drapeau had successfully lobbied to hold the fair in his city, but where? Numerous studies of sites across Montreal and the surrounding area were conducted, but there were huge problems with each of them. No single site met the requirements of size, availability, and public transportation. Some consideration was given to using multiple sites, somehow linked together, but on March 22, 1963, the organizers announced that they had found a rather unusual solution to their problem. They would build their fair in the middle of the St. Lawrence River.

What was to become Expo 67 was originally two islands, Île Sainte-Hélène and Île Ronde, and a section of the city's port facilities. These areas were not nearly large enough to hold all the pavilions planned for the fair, even if they were expanded, so a completely new island, Île Notre-Dame, was built. The decision to build the islands had not been a unanimous one, and many of the corporation's leaders resigned in protest, some stating that the project was too ambitious, too costly, and simply could not be done on time. Drapeau refused to listen to their complaints and pressed on, managing to fill the board with members more sympathetic to his plans. The City of Saint-Lambert was also not pleased with the plan, claiming that section of the river was within its boundaries, and it took a federal ruling to resolve the dispute in Montreal's favor.

Work on building the islands began on August 13, 1963. An estimated 28 million metric tons of rock and dirt were needed to build out the site, much of it coming from the excavations for the new Montreal subway system, expansion of the St. Lawrence Seaway, and by dredging the river near the construction site. When completed, the site was slightly over 1,000 acres.

During construction of the islands, the world's fair had been rebranded as Expo 67. The name was suggested by Donald A. Logan, a BIE executive, who noted that it would help make the event stand out from other recent world's fairs and was easily understandable in both French and English. The planners had also chosen the theme "Man and His World," based on the 1939 book *Terre des Hommes* by Antoine de Saint-Exupéry. An Expo press release explained that "the theme 'Man and his World' (would) provide a stirring illustration of 20th Century humanism in a fully integrated presentation of Man's artistic, scientific and philosophical development, the whole permeated by both a feeling of belonging to the community of Man and an awareness of the basic unity of mankind." Most of the pavilions and exhibits found some way, even if tenuous, to link to the overall theme.

The islands were turned over to the Exposition corporation on June 30, 1964, and construction work on pavilions and supporting infrastructure began in earnest. Building Expo 67 was a monumental job, employing more than 6,000 workers, but it was done on time. When completed, the site held 847 buildings and pavilions; 27 bridges; 51 miles of road and walkways; 23 miles of sewers and drains; 100 miles of water, gas, and power lines; 55,000 miles of phone wires and cables; 24,484 parking spaces; 14,950 trees; 4,330 trash cans; and 6,150 light fixtures.

Expo 67 opened on April 28, 1967. A total of 50,306,648 customers enjoyed the fanciful pavilions and shows before the gates closed on October 29, making it one of the best-attended expositions in history. That was not the end of Expo 67, though, as seen in chapter 5.

Like most world's fairs, Expo 67 ran over budget. The 1963 bill that authorized the fair was based on a budget of $167 million. Costs soared to $218 million, which would be $1.5 billion today if adjusted for inflation. Some reports claim the final sum was actually $439 million if all of the infrastructure and operating costs were included. Those who attended Expo 67 are likely to agree the money was well spent, though, as the summer of 1967 was a wonderful time to have been in Montreal and part of Expo. Many modern-day residents and visitors to Montreal would also agree the project was worth it, for most of the former Expo grounds are now Parc Jean-Drapeau, a beautiful oasis that is a welcome addition to the city.

# *One*

# CITÉ DU HAVRE

The area chosen for the main entrance to Expo 67 was Mackay Pier, a man-made peninsula originally built to protect the port from storms. Before construction for Expo began, it was a little-used spit of land that was largely undeveloped; this early photograph shows the area after it was cleared, enlarged, and renamed for Expo. The white oval structure would become the Automotive Stadium and was the site of many events during Expo.

There were many ways to get to Expo 67, including private car, bus, taxi, the Metro system, or by foot. One of the most novel ways was via hovercraft, which connected several stops on the site to downtown Montreal. Many riders found the hovercraft a bit unsettling; the waters of the St. Lawrence River, especially in the LeMoyne Channel between the exposition's islands, made for a bumpy and rather uncomfortable ride.

Seven towering hexagons marked the Place d'Accueil, the main entrance to Expo 67. The bustling 2.5-acre plaza featured a stop on the Expo Express train line, a bus depot, a large taxi facility, and an Expo tram that took riders farther into the grounds. Information counters were available for those looking for accommodations or other tourist services. There was also a wide variety of shops, restaurants, lounges, a day hotel, and even a liquor store.

Expo 67 covered 1,000 acres, with some very long bridges that connected the islands, making it impossible to see it all in one visit. Here, a group of visitors studies a map just inside the entrance as they plan out their day. There were also numerous information stands to help guests find their way around the grounds.

Visitors wishing to get quickly to the main islands of Expo 67 could board the Expo Express train at Place d'Accueil, then travel to stops at Habitat, Île Sainte-Hélène, Île Notre-Dame, and the La Ronde amusement area. The 3.5-mile route afforded excellent views of the grounds and proved very popular with riders. The free trains were completely automated, making it the first such system in North America, but operators were placed on board to allay the fears of nervous passengers.

Expo 67 hosted the largest art display of any world's fair. In addition to dozens of sculptures scattered across the site and artwork within many of the international pavilions, there were 188 paintings from 40 countries and 50 sculptures on display. A special $35 million insurance policy covered the exhibits inside the 20,000-square-foot Art Gallery, which was designed to exacting art museum standards.

This large bronze sculpture was one of the many pieces of artwork created for Expo 67. *Tall Couple* by Louis Archambault was commissioned by the House of Seagram and was prominently located outside the Art Gallery and on the main route from the Place d'Accueil to the rest of the pavilions. It is now located at the University of Toronto, Scarborough campus.

The Photography and Industrial Design pavilion combined two different themes within its walls under the umbrella theme of "Man the Creator." "The Camera as Witness" presented 500 photographs from around the world, covering a wide variety of subjects and styles. Industrial designs from 18 major schools comprised the second half of the exhibit. Many visitors delighted in taking unusual photographs of the area using this polished metal globe.

Labyrinth was the home of an impressive film project sponsored by the National Film Board of Canada. Loosely based on the ancient Greek fable of Theseus, who slew the Minotaur in its maze, Labyrinth told the story of how mankind was making its way through an even more elaborate maze as people progressed from living in caves to their modern lifestyles. New projection techniques included dual 70-millimeter projections on two perpendicular screens, which is now credited as being the forerunner to the popular IMAX format.

Habitat 67, also known just as Habitat, was a very unusual exhibit, as people actually lived in the complex while Expo 67 was open. A total of 354 preformed concrete cubes were used to construct the 158 apartments in the striking complex. The apartments came in a wide variety of configurations, ranging from a simple single-bedroom unit of 600 square feet to a four-bedroom complex of 1,700 square feet. Each precast cube measured 17 feet, 6 inches by 38 feet 6 inches by 10 feet high and weighed 70–90 tons.

Although the exterior of Habitat was quite unusual, the interior spaces were rather conventional. Living inside an ongoing exhibit had some unusual challenges. Expo visitors could take tours of a fully furnished sample apartment; when the line got too long, some tourists resorted to knocking on the doors of the occupied units to ask if they could take a quick peek inside. Even worse, any friends coming for a visit had to pay the Expo 67 admission fee.

Guests could also tour the specialized equipment used to build Habitat, including the $750,000 cranes, concrete molds for the cubes, and a half-finished unit that showcased the interior construction techniques. Habitat was originally planned to be a 900-unit complex, and others were to have been built around the world. Higher than expected costs led to a scaling back of the project, and none of the other complexes were ever built. Although generally panned by critics in 1967, Habitat has become one of the more exclusive addresses in Montreal.

A dock opposite Habitat provided the opportunity to tour a number of ships while they were visiting Expo. Many were Canadian warships, such as the destroyer HMCS *Restigouche* seen here, but others came from 11 different countries. The sailing ship *Bluenose II*, which was designated the host ship of Expo, was also on display.

The soaring wooden spire of the Man in the Community pavilion was a latticework structure made of wood from the Pacific coast. It was open to the elements, and rainwater could pass through into a large pool inside the main building. Surrounding the pool were displays on mankind's propensity to reshape its environment and to live together in large cities. Some of the displays noted the unfortunate environmental decline that often results from poorly planned projects and careless inhabitants.

The 2,000-seat Expo Théâtre was located just outside the main gate, which allowed patrons to enter without paying an admission fee for Expo itself; evening performances included free admission to Expo. In addition to appearances by well-known celebrities, there were films sponsored by the Montreal International Film Festival. As part of the World Festival program held in Montreal that year, the theater also hosted a wide variety of performers from around the world.

"Man and Music" was the theme of the Jeunesses Musicales of Canada pavilion, which was sponsored by the Portland Cement Association. Visitors could listen in on sessions that demonstrated different methods used to teach music, or follow the history of music in Canada since 1610. The pavilion was also the home of the 21st World Congress of the International Federation of the Jeunesses Musicales. It is now part of the organization's camp in Mount Orford, Quebec.

The Canadian Broadcasting Corporation (CBC) spent $10 million to construct the International Broadcasting Centre, which included two television and six radio studios, as well as editing suites, control rooms, and equipment repair facilities. In addition to the CBC, more than 200 other news organizations used the facility to file stories on Expo 67. Tours were given of the facility, and guests could relax while watching CBC programs in air-conditioned comfort.

More than 100 companies joined forces with the Quebec Department of Industry and Commerce to sponsor the Quebec Industries pavilion. As might be expected, the focus was on attracting new customers to the region. Many of the exhibits touted the availability of freshwater supplies, including live video links to Hydro Quebec's massive new power dam on the Manicouagan River.

A bronze statue of famed Finnish runner Paavo Nurmi stood outside the entrance to Olympic House, which celebrated the theme of "Man at Play." Inside were exhibits on the Olympic Games, with an emphasis on Canadian participation and achievement. Many of the other international Olympic committees also had smaller displays. Sadly, the statue was damaged when it was toppled by vandals; it was repaired during Expo and is on display now in Helsinki.

The Yukon Territory had the smallest of the Canadian pavilions and was almost not at Expo at all. For financial reasons, the government had initially declined to participate, and when it finally decided to be part of the event, the only space left that was within its budget was in a string of small shops. The tiny exhibit, located near the red van in the photograph, proved to be a popular stop for those wishing to learn more about the nation's smallest territory.

One of the largest buildings on Cité du Havre was the Administration and Press pavilion, which was not open for public tours. In addition to office space, it contained facilities for press conferences, meeting rooms, and interpreter services. One of the few buildings designed to be reused after Expo closed, it is the home today of the Montreal Port Authority and the Maritime Employers Association.

An exposition full of futuristic pavilions needed a futuristic infrastructure. This included a novel design for Expo 67's phone booths, each of which was topped by a large plastic dome. The design proved less than successful; the domes did not keep callers dry in the rain, and on sunny days they focused the heat down towards the phones. They also seemed to magnify passing sounds, or those of other callers. It is little wonder they were never used again elsewhere.

The unusual-looking poles topped with circular discs set at rakish angles were Expo's street lights. Rather than using lights on top of the poles like conventional lighting, the Expo units housed the bulbs in the bases, reflecting them off the discs to softly illuminate the surrounding area. The colored structure on the left held the speakers for a small bandstand.

# Two

# ÎLE SAINTE-HÉLÈNE

After enjoying the sights at Cité du Havre, guests would make their way to Île Sainte-Hélène and the rest of Expo across the St. Lawrence River via the Concordia Bridge. The 2,265-foot-long bridge was unusually wide at 94 feet and was designed to carry pedestrian, vehicle, and train traffic. This view from Habitat shows how swiftly the river moves at this spot; the builders overcame the 12-knot current by prefabricating the bridge sections on shore.

The Place des Nations was the site of the almost daily ceremonies celebrating the participating nations, Canadian provinces, and other special groups that had traveled to Expo 67. The 7,000-seat arena also featured an Olympic-style torch, seen on the left, which was kept burning throughout the duration of Expo, as well as flags of the international participants.

A wide variety of groups came from across Canada and the United States to perform at Expo 67. Inviting these groups was a masterful marketing strategy; many of the families of the performers also made the trip to Montreal, thereby adding to the number of paying customers. Here, students from the Kuper Island Indian School in British Columbia perform at Place des Nations.

One of the most beautiful areas in Expo 67 was Swan Lake, which was next to the Expo Express station. Powerful jets of water arced into the air, at times providing a welcome mist on hot days. Several families of swans were moved there to help legitimize the lake's name. Facing the lake were the tetrahedral-shaped Man the Explorer theme pavilions; the geodesic dome of the United States is seen in the rear.

The International Nickel Company of Canada Ltd. sponsored Alexander Calder's massive sculpture *Man, Three Disks*, his largest work. Built of 48 tons of stainless steel, the sculpture reached nearly 80 feet high above its concrete base, which housed a lounge, flower beds, and smaller pieces of art. *Man, Three Disks* has been moved from its original site next to Swan Lake but remains in the park built on the former Expo site.

The exterior of the Netherlands pavilion was crisscrossed by 35 miles of aluminum tubing and a surrounding terrace of Dutch bricks. The displays inside showed how the country overcame its battles with the sea outside its dikes and became an economic powerhouse. Examples of modern Dutch industries were shown alongside exhibits of life in the nation's far-flung colonies.

Native materials were used to construct the Belgium pavilion, a striking building with a modern-looking mix of glass, wood, and brick. Inside was a wide variety of exhibits that ranged from famous artworks to examples of modern products and industries. One section showed how Belgium had profited from Expo 58, while another took a humorous look at Tintin, the famous Belgian comic book character. The pavilion also housed Le Bruxelles, one of Expo 67's most popular restaurants.

Despite being landlocked and possessing limited natural resources, Switzerland has become a major economic power. Static displays and a 20-minute film showed the transformation from a sleepy farming society to the modern cities of the time. There was, naturally, a focus on tourism and banking. Four restaurants, one for each of the country's regions, offered their local foods and wines.

A master clock in the Switzerland pavilion was linked to 12 others on the site to provide the official Expo 67 time. Visitors could observe the master clock, which was advertised as being accurate to a millionth of a second per day, as part of a display on Switzerland's famed watch and clock industry.

The sloping roofs of the Austria pavilion were said to resemble crystal forms that suggested thoughts of "mountains, precious stones, a romantic countryside, scientific precision, and achievement in the arts." Inside was a multimedia show dubbed "Austrovision," which showed examples of life in Austria. The pavilion may best be remembered, though, for its popular Wienerwald restaurant, which offered a comfortable spot for diners to enjoy the views.

One architect from each of the five Scandinavian countries—Denmark, Finland, Iceland, Norway, and Sweden—designed a section of this stark but elegant pavilion, which was constructed almost entirely of materials imported from Scandinavia. The building included a sculpture garden; upstairs was a restaurant as well as exhibits that highlighted life in each of the countries.

The Vienna Kindergarten allowed parents to enroll their children, ages three to six, in a two-week Montessori kindergarten program conducted in this colorful pavilion. Students were also admitted daily on a space available basis for either half-day or full-day sessions. The parents were free to enjoy Expo on their own, or they could watch the children from outside the fence surrounding the play area.

Rather than starting at the bottom and working their way up, as at most pavilions, visitors to Japan rode to the top floor on escalators then toured the exhibits on their way back to the ground level. In addition to displays about Japan's culture, business, and tourism, there was a preview of Expo 70, the next world's fair scheduled for Osaka. A traditional Japanese garden provided a quiet oasis.

The Man the Explorer complex consisted of three buildings shaped like truncated tetrahedrons arranged around the Plaza of the Universe, which featured an open courtyard and restaurant. Connected by steel bridges, the buildings housed four exhibits: Man and Life, Man and the Oceans, Man and the Polar Regions, and Man, his Planet and Space. The electronic information sign was quite an innovation for 1967; several were located around the grounds and were updated with schedule messages throughout the day.

Man and the Polar Regions explored the frozen worlds of the Arctic and Antarctic. Visitors entered through a chilled tunnel and were treated to a vista of frozen wastes that were apparently free of life. Displays and an 18-minute film showed that there was actually an abundance of life at the poles, including hardy residents who had learned how to deal with the harsh conditions.

The Green Mountains of Vermont were the inspiration for the soaring rooflines of the state's pavilion. One of the highlights inside was watching sculptor Ferdinand L. Weber as he carved a piece of Vermont granite into a 12-foot-tall statue of Samuel de Champlain, the founder of Quebec and namesake of the 125-mile-long lake that stretches from the Canadian border deep into Vermont.

The Brewers Pavilion had something for everybody. Children were treated to 15-minute bilingual shows by the Canadian Puppet Theater, while adults could choose from 60 different brands of beer. La Brasserie, the pavilion's restaurant, was one of the most highly rated food outlets at Expo, receiving rave critical reviews for both the food and friendly service. On the way out, there were plenty of free cookbooks featuring recipes made with beer.

Beautiful ceramic tiles arranged to look like rolls of Persian carpets covered the exterior of Iran's pavilion. Numerous models, dioramas, films, and photo montages were used to tell the story of Iran, from the ancient days of its founding to modern life under the shah. Viewers could watch live stage shows while they enjoyed Iranian vodka, caviar, and other foods, or study rug weavers at work.

Walt Disney's innovative 22-minute film *Canada 67*, presented on nine screens in Circle-Vision 360, was the star of the Telephone Pavilion. Other exhibits allowed children to call Disney characters, let guests experience the new Picturephone service, try to beat a computer at tic-tac-toe, or have to their ages guessed electronically. The history of telephone service in Canada was on display, along with a live show explaining how telephones would soon be used for electronic shopping and banking.

Perhaps it is ironic, but the Polymer pavilion, sponsored by a plastics company, was actually built out of concrete. Resembling a giant wheel when viewed from overhead, it held exhibits along its spokes that were designed to reveal why mankind is so curious and eager to learn. The multilevel structure was an intriguing maze of curving ramps, hidden rooms, and reflecting pools. At the top, an 18-foot-tall metal mobile represented a polymer molecule.

Air Canada reached back into history to the designs of Leonardo da Vinci for its pavilion. The helix-shaped structure featured 23 cantilevered blades that resembled one of his concepts for a "flying machine" as well as suggesting the blades of a modern jet engine. Displays followed mankind's pursuit of flight, from balloons and early gliders through to the latest airliners, with an emphasis on how easier travel has blurred social and governmental boundaries.

The colonial facade of the Maine pavilion was an eye-catching contrast to its futuristic looking neighbors. As visitors approached it, they could spot a model train carrying the state's products circling the building; if they followed it inside, they were treated to exhibits on Maine's industrial capacity, fishing fleets, and year-round tourism opportunities. A simulated gristmill housed a 15-minute film extolling the state's merits.

Sharing a common border with Canada, the State of New York was eager to sell itself to potential business partners and tourists. Many of the displays touted how well Canadian companies were doing in New York, with trade agents on hand to explain why others should look to New York for both customers and raw materials. Other exhibits under the six stylized mountains showcased the natural beauty of the state's many diverse geographic regions.

A colorful yellow roof marked the International Scout Centre, which represented the 11 million members of Scouting programs in 89 participating nations. Visitors were invited to watch demonstrations of Scout activities that included bridge building, cooking, first aid, and in a pool built specially for the occasion, water-based skills such as swimming, scuba diving, and lifesaving. Scouts from Canada and the United States served in the Scout Service Corps, which provided assistance to visitors across all of Expo.

The Republic of Korea exhibited in a pavilion that was notable for being built almost entirely of wood. The exhibit hall combined ancient Korean art with samples of the latest industrial goods manufactured in Korea's modern factories. The building is one of the last Expo structures still in place but is in disrepair today and no longer used. Its once soaring tower lies toppled on the ground.

A devastating fire caused by a dropped cigarette destroyed the exhibits inside the Republic of China's pavilion on May 30. A massive effort by the Taiwanese government to repair the damage resulted in its reopening on July 2. A popular Chinese restaurant located in a smaller building behind the damaged main hall was able to reopen for business during the repairs.

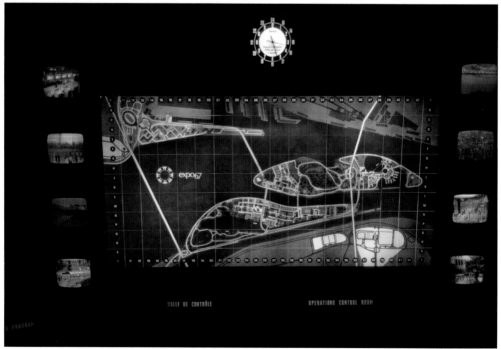

Guests were invited inside the Operations Control Center for a look at how Expo 67 was run. Sponsored by the Canadian General Electric Company, the displays proclaimed how computers and other modern technology were used to keep things flowing smoothly. The large map at the center of the room, ostensibly showcasing activity across the site, was actually just a static display, but the television monitors provided real-time views from across the grounds.

One of the best ways to tour Expo 67 was aboard one of the Minirails, automated slow-moving monorail systems that passed between the pavilions. The open-air cars provided excellent views of the site, far above the often crowded streets. The system seen here circled through the Île Sainte-Hélène; another serviced La Ronde, while the longest system linked Île Sainte-Hélène and Île Notre-Dame.

One of the Minirails actually passed through the United States pavilion. At 187 feet high and 250 feet in diameter, the United States pavilion was the largest transparent structure in the world. The geodesic dome was comprised of 1,900 plastic panels attached to a tubular frame. The unusual design made it one of the most eye-catching pavilions at Expo and a favorite for photographers.

The largest display area in the United States pavilion featured a collection of American spacecraft including manned capsules and satellites. Some of the ships had actually flown into space, such as *Freedom 7*, the Mercury capsule flown by Alan Shepard, the first American in space, and *Gemini 7*. There was even an early Apollo capsule used in a suborbital flight in preparation for an eventual flight to the moon.

The United States pavilion was one of the most popular at Expo; very long lines stretched far outside as people waited for a chance to get inside. When they finally got in, visitors began their tour with a ride on the world's longest escalator. They were then treated to a wide variety of exhibits that included 300 types of hats worn by American males, photographs of Hollywood stars, presidential election campaign posters, and a giant collection of Raggedy Ann and Andy dolls.

There was much more to Île Sainte-Hélène than the fanciful pavilions on the west end or La Ronde on the east. The center of the island, which was the section originally there before the massive landfill operation that shaped the site for Expo, was a restful oasis of gardens and tree-lined paths. Many guests found a relaxing moment in the International Rose Garden, a critically acclaimed botanical treasure that featured 10,000 plants from 15 countries, dotted with dozens of sculptures.

Guests could travel across the vast expanses of Expo aboard one of the La Balade trailer trains that connected Cité du Havre and Île Sainte-Hélène. One is seen passing the Hélène de Champlain Restaurant, which originally opened in 1938 and was built from locally quarried stone. During Expo, it was reserved for official functions and VIP visits, but it is now, once again, open to the public.

Lévis Tower, a well-known landmark on the highest point in the grounds, was capped by the Sun Life Centenary Carillon, the largest electronic carillon in the world. The 671 bells could be heard for miles and were used to start and end each day with the Expo theme song, "Hey Friend, Say Friend." A glass-enclosed booth allowed visitors to watch the carilloneurs as they played the instrument; the carillon was also automated and could play preprogrammed tunes.

Unnoticed by most visitors, this bland-looking building served a vital purpose. This was the Public Safety Building, which housed security and medical services, including a branch of the St. John's Ambulance Brigade. Additional first aid stations were located throughout Expo. Happily, while the emergency teams were kept busy with a stream of minor injuries and other issues, there were no major health or safety issues during Expo's run.

*Three*

# ÎLE NOTRE-DAME

While each of the other sections of Expo 67 were all quite interesting on their own, the real heart of the exhibition was on the man-made island Île Notre-Dame. The island held the greatest concentration of exhibits, including the largest one (the Canada pavilions complex) and two of the most expensive ones (USSR and Czechoslovakia). Although almost every inch of available land was covered by a pavilion, an excellent transportation network that included wide walkways, the Minirail trains, and canal boats helped make it easy to tour the island.

The Bridge of the Isles was one of three spans over the LeMoyne Channel between Île Sainte-Hélène and Île Notre Dame. The designers of the reinforced concrete and steel bridge were told to put the support towers and cables in the middle of the span for aesthetic purposes, which led to premature wear and cracking. When the bridge was refurbished years later, new piers were added underneath the bridge and the cables were removed.

Cosmos Walk was a pedestrian walkway that also carried one of the Minirail lines that connected the islands. Seen here with one of the hovercraft speeding underneath, the bridge's name was, at least in part, due to the large space exploration displays located in the United States and USSR pavilions at either end. Often jammed with guests, Cosmos Walk was usually the most crowded part of Expo.

Riders arriving at Île Notre-Dame on the Expo Express train could disembark at the Transportation Plaza station or continue on to La Ronde. The plaza was a major stop on the Minirail line, and many visitors began exploring the island with a ride on the system. There were so many riders, in fact, that Expo authorities had to institute a new fare plan to force people off the ride at some stops in order to make room for others.

There were three separate Minirail lines. The blue-and-white cars seen here were used on the 4.2-mile segment that linked the islands and ran throughout Île Notre-Dame. That Minirail was designed in Switzerland, with the cars and track built in Canada. The yellow-topped Minirails that serviced Île Sainte-Hélène and La Ronde also came from Switzerland and had been purchased secondhand at the end of the Swiss National Exhibition in 1964.

As it was the host nation, it would seem natural that Canada would have had the largest exhibit at Expo 67, but the initial site plan only allocated one acre for the pavilion. Architect Rod Robbie pushed successfully to get the largest plot on the grounds for Canada, an impressive 11.5-acre complex of buildings and open space. Seen here from the France pavilion, the Canada pavilion included a 1,200-seat open-air amphitheater where 58 performances were scheduled each week.

The Canada pavilion featured 125 exhibits under a variety of pyramid-shaped roofs that were inspired by the nation's minerals and metals. The displays covered a wide gamut of subjects: the Land of Canada, the People of Canada, the Growth of Canada, the Challenges to Canadians, and Canada and the World. There was also an arts center, a section for meditation, a children's playground, and two restaurants, the Tundra and the Buffet, serving popular dishes from across the country.

The largest structure in the $24 million Canada pavilion was the Katimavik, a giant inverted pyramid named after the Inuit word meaning "gathering place." The structure was so unique that it was chosen as the image representing Expo 67 on a Canadian postage stamp. Standing 109 feet high, the Katimavik dominated the skyline at the end of Île Notre-Dame and could easily be seen from downtown Montreal.

The Katimavik was even more unusual on the inside. Visitors could climb staircases to a walkway that provided spectacular views of Expo and Montreal. The walls were dotted with numerous sculptures, many of them kinetic pieces that were synchronized to eerie electronic music. A funicular elevator provided access to the top for those who did not want to trek up the stairs, but inadequate air-conditioning inside led to many riders deciding to walk back down.

Canada's new maple leaf flag was the inspiration for the People Tree, a 60-foot-tall stylized maple tree comprised of 1,500 red and orange three-by-five-foot nylon sheets. Approximately 700 of the panels featured images of Canadians at work and play. The photographs were selected by the National Film Board of Canada from archival sources and five commissioned photographers. Recordings of sounds from across Canada were intended to suggest the passage of wind through the "tree."

A $65,000 fleet of radio-controlled ships set in a 150-by-300-foot lagoon entertained visitors in a show sponsored by the Canadian Coast Guard and the Department of Transport. Two icebreakers came to the rescue of a massive ore carrier trapped in an ice jam, then safely escorted it back to the dock. Although the ice pack was actually just chunks of Styrofoam, the 20-minute show was a hit with both adults and children.

The pyramid motif of the Canada pavilion was carried over to the Ontario pavilion, which featured a fiberglass membrane roof stretched over steel supports. Underneath were hundreds of massive granite blocks that reflected the strength of the province and Canada. They were popular with children who liked to climb on them, but they undoubtedly led to many nervous moments for their parents. The pavilion featured the 17-minute film *A Place to Stand*, an Academy Award winner for Best Live Action Short Subject.

The mirrored glass walls of the Quebec pavilion gave it two completely different looks. During the day, the slightly sloping walls of the truncated pyramid reflected the sky, making the building seemingly change shape as clouds passed overhead. At night, the walls allowed the interior lights to shine through, producing a 50-foot-tall block of light that looked like it floated above the water due to its unlit pedestal base.

Gerald Gladstone's *Uki* was a unique kinetic sculpture. For much of the time, it looked like little more than some rusting shapes protruding from the waters behind the Canadian pavilion. Then, water cascaded off what was revealed to be the welded figure of a stylized dragon. Its two heads emerged every 30 minutes to let loose a bellowing blast of flame before the beast hid again. *Uki* is a Huron-Iroquoian term for monster.

Nova Scotia, New Brunswick, Prince Edward Island, and Newfoundland shared the Atlantic Provinces pavilion. Most of the exhibits were outdoors under the shelter of the world's largest cantilevered wooden roof. Exhibits reflecting the close bond between these provinces and the sea were capped by the construction of a 47-foot schooner, the *Atlantica*, in front of the building; it was launched on October 11, 1967. The pavilion's seafood restaurant was very popular, with five-hour waits quite common.

Manitoba, Saskatchewan, Alberta, and British Columbia were represented in the Western Provinces pavilion, a wooden-roofed building capped by giant Douglas firs protruding through the roof. Inside, visitors were treated to displays about life and work in the west; the most elaborate exhibit was a simulated trip down 3,000 feet inside a mine. Other displays included massive logging trucks; the sights, smells, and sounds of forests and grain fields; and a look at how the area has changed since it was first settled by European traders.

In 1967, asbestos was widely used in building construction, and trade advertisements proudly proclaimed how the mineral was used in many Expo 67 pavilions. The industry was saluted at Asbestos Plaza, which featured a huge piece of asbestos ore atop a six-foot-high fountain designed to make it look like the rock was floating atop the water jets. Even the flower planters and benches were made with asbestos. In just a few more years, the carcinogenic properties of asbestos became known and most uses were discontinued.

A 100-foot-tall teepee was the centerpiece for the Indians of Canada pavilion. The pavilion was one of the more controversial exhibits at Expo 67 due to a number of displays highlighting some of the broken treaties with the Indians over the years and a plea to "give us the right to manage our own affairs." The pavilion also featured brightly colored murals by artist Francis Kagigewikwenikong and a 65-foot-high Kwakiutl-style totem pole carved by Tony and Harry Hunt.

The flags of its 122 member nations flew outside the United Nations pavilion. Inside was a 330-seat theater showing *To Be Alive*, an Academy Award–winning film first shown at the 1964–1965 New York World's Fair. A post office sold special commemorative stamps, and the Restaurant of All Nations offered selections of foods from around the world. Outside, the *Tree of Life* sculpture, carved into a 150-year old chestnut tree, showcased faces from five continents.

The Jamaica pavilion was based on a 19th-century Jamaican country store, a traditional shopping venue that also included a tavern. The real country stores had used a variety of architectural styles borrowed from the nations that traded with Jamaica, and the Expo copy faithfully reproduced many of these elements. The building was completely renovated in 2008 and is one of the few Expo structures still in use today.

The principality of Monaco featured a unique theater that was as much a garden as it was a spot to watch a film. Hundreds of colorful blooms surrounded viewers as they watched a 15-minute film highlighting the history, tourism opportunities, and the varied businesses of the tiny Mediterranean country. Other displays included information on the nation's famous Grand Prix car race and on its principal export, colorful postage stamps.

Haiti's pavilion was simple in design but still quite attractive. Large walls of glass beads provided a tantalizing peek at the displays inside, and the aromas emanating from the Tierra del Soleil restaurant further lured guests inside. There, they were treated to displays featuring Haitian art and products made from native cloths and fibers. A large outdoor dining area on the water made it a pleasant spot to dine and to relax.

Three professional engineering societies funded Engineers' Plaza, an open-air tribute to the work of engineers from across Canada. At the center was *Space Column*, a 40-foot tall sculpture by Gerald Gladstone. The sculpture and surrounding fountain, made of welded and riveted aluminum, represented a galaxy with planets and stars orbiting in space. Gladstone was the only artist to receive three commissions for artwork to be displayed at Expo 67.

The theme of the Christian pavilion was "The Eighth Day," when God is said to have rested after creating the world and mankind was free to start shaping the planet to suit its needs and dreams. Sponsored by eight Christian denominations, the pavilion was controversial due to many disturbing scenes in a 13-minute film of wartime atrocities, drug abuse, murders, and other moments where people turned away from Christ.

Greece had one of the starkest pavilions at Expo 67, comprised of several windowless buildings finished in plain white walls. The only splash of color came from the plants surrounding the site, which were all imported from Greece, and from the costumes of performers entertaining in an open-air courtyard. Displays inside the halls traced Greek contributions to art, religion, science, sports, and modern technologies, and showcased tourism and business opportunities.

The tiny island nation of Mauritius was celebrating its newly granted independence from Great Britain with exhibits touting its important sugar, tea, and tobacco industries. There was only one exhibit room inside the 55-foot-tall nylon-roofed structure, but it was full of models, photographs, and detailed topographic maps of the farm areas. Hostesses were available to explain the country's culture and arts as native music played in the background.

The Mauritius pavilion was set next to one of the canals that ran through Île Notre-Dame. Several types of boats were used to ferry passengers, including the Vaporetto motor launch seen here, which were named after water-buses in Venice. There were also motorized gondolas, pontoon boats, and other small craft. Some ran on a scheduled basis while others were hired by the hour for more relaxing sightseeing tours.

The 21 member nations of the Organization for Economic Cooperation and Development sponsored a pavilion to explain the tenets of the organization, which was founded in 1961 to promote economic growth and stability through increased international trade. Since Expo 67, the Paris-based trade group has grown to 34 countries. Sharp-eyed readers may note there are 23 flags on the building; two were additional countries considering membership.

Seven soaring triangles marked the Yugoslavia pavilion, a design that produced a surprisingly large amount of exhibit space. One of the largest displays was on Yugoslavian art and history; others showcased the business and tourism aspects of the country. One large section saluted the nation's diversity by noting that it had "one state, two alphabets, three languages, four religions, five nationalities and six republics."

On display inside the Israel pavilion was a complete Dead Sea Scroll, part of a display about the antiquities and history of the region. There was a more modern side as well, with exhibits on the Holocaust and the many difficulties faced by the still young nation. Modern Israeli crafts and products were on display, and the pavilion featured the only kosher restaurant at Expo 67. Live performances and films were shown in a 250-seat theater.

The Trinidad & Tobago and Grenada pavilion, perhaps the longest-named exhibit at Expo 67, looked at some of the common features of the two island nations, which had strong ties to Great Britain, but it also showed some of the differences between them. The pavilion may be best remembered, though, for the lively steel band performances held on a small stage that jutted out into the large pool fronting the building.

Another one of Expo 67's theme pavilions was located on Île Notre-Dame. Man the Producer had three sections: Resources for Man, Progress, and Man in Control? All were focused on telling the story of how mankind has used the planet's abundant natural resources to improve their lives but sometimes at a heavy ecological cost that may negatively impact future generations.

A curved roof sweeping gracefully upward marked Italy's participation at Expo 67. It was unusual in that it featured three sculptures on the roof that represented the pavilion's themes of "Tradition, Customs, and Progress." The most memorable of those was the burnished bronze globe by Arnaldo Pomodoro, which glinted in the sun and drew attention to the structure. Inside were exhibits of Italian antiquities and the latest scientific advances.

A 200-foot tower, the tallest at Expo 67 and topped by a stylized Union Jack, marked the location of the Britain pavilion. Upon entering the building, guests were transported back in time to the earliest days of Great Britain, then treated to displays showing how the country had grown to become a world leader in commerce, the arts, and technology. The final section combined predictions for Britain's future with a plea for world peace.

The Swinging 60s were in full force, so it was inevitable that music by the Beatles and the Rolling Stones would be heard playing in the background as guests toured the Britain pavilion. The mod look was extended to the colorful costumes worn by the pavilion's hostesses. There were traditional elements as well—the background music also included snippets from Gilbert and Sullivan—but the focus was very much on the Britain of 1967 and of the future.

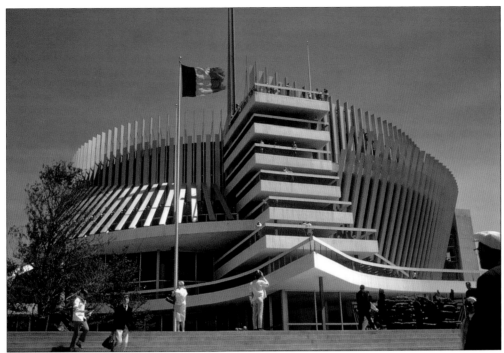

Large aluminum ribs were used to make the already impressive-looking France pavilion seem even larger than it actually was and to reduce sunlight on the glass-walled pavilion. The concrete building housed exhibits on eight different floors, with a sculpture garden on the roof. Not content with all of that, and to compete with the nearby Britain tower, the designers added a massive aluminum arrow that soared high into the sky, making it easy to spot the pavilion, especially at night.

Inside, the exhibits were spread throughout exhibition halls that circled the open-air center of the 220,000-square-foot building. The latest French fashions were there, as well as samples of French technology, but for many visitors the highlight of the exhibits was a well-received collection of artwork on loan from museums across the country. Auguste Rodin's *Eve*, which took him 18 years to finish, had a prominent location at the top of one of the building's major staircases.

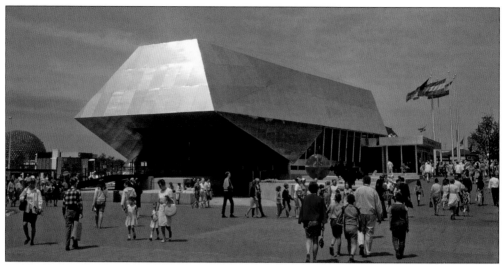

The European Communities pavilion was presented by the European Common Market (a predecessor of today's European Union) and two trade groups to promote their goal of improving economic growth by removing trade barriers between the six member nations. The highly polished steel surface was intended to represent the strength of a diamond, inferring the strength the union brought to its members. The mirror-like pavilion proved to be hard to photograph due to the sun's glare at most angles.

The Canadian National pavilion challenged visitors to reconsider their concepts of time and motion through exhibits and a film that demonstrated how people may perceive the same event differently. The pavilion consisted of nine polyhedrons; some were merely open frames while others held exhibits viewed from a garden area underneath. The complex also included an eight-sided, 200-seat theater.

Some days at Expo 67 were quite crowded, as seen in this July 1967 photograph. It is likely that the people in line for the Canadian National pavilion (left) had a new appreciation for its exhibits on Time and Motion when they finally got inside. The neighboring pavilions were all also very popular, as seen by the long lines for Kodak (top) and Cuba (bottom right). The crowds were generally well behaved, with very little report of problems due to the long waits.

Amateur photographers were invited to learn how to take better photographs using tips from the staff at the Kodak pavilion. Visitors could also get pamphlets showing preselected locations for the best views of Expo 67, buy film, or get their cameras cleaned and adjusted. The pavilion also featured a unique slide show presentation, *Wonders of Photography*, that used 12 projectors, and, in a well-received finale, a "screen" comprised of water mists that imparted a 3-D feel to the images.

Cuba had the most politically oriented pavilion at Expo 67, a fact that was not lost on the press or the many visitors who found the exhibits disturbing. The pavilion showed the differences in Cuba before and after the 1959 revolution, with graphic depictions of violence and oppression to help make its argument that the revolt was justified. The displays and hostesses hailed the efforts of Prime Minister Fidel Castro in changing the nation, but the hard sell approach failed to win many fans.

The Chatelaine House was sponsored by *Chatelaine* magazine and the Canadian Lumberman's Association. It was touted as a modern way to live in a small space, with the latest space-saving appliances and an artificial lawn. The house and its contents, including the car in the driveway, were to be given away to a lucky guest and moved to a new site in Ottawa. The winner was Douglas McEachen, a 17-year-old from Regina, who opted instead for the alternate prize of $30,000.

Czechoslovakia was the surprise hit of Expo 67, attracting such large crowds that the line waiting to get in often stretched past several of the neighboring buildings. It was so popular that former American first lady Jacqueline Kennedy made it the first stop during her three-day visit to Expo in October. The pavilion cost an astonishing $10 million, just behind the $15 million USSR building and even more than the $9 million United States dome.

There were many different exhibits inside the massive structure. One of the most popular was an expansive display of glassworks, some in the form of sculptures and others in more prosaic form, such as dinnerware. Even the ashtrays in the pavilion's popular restaurant were made of blown glass. Carefully placed lighting brought many of the glass pieces to life, making a nighttime visit especially enjoyable.

Multimedia presentations played an important role in the Czechoslovakia pavilion. One of the more unusual was *Diapolyecran*, a slide show that projected 15,000 images on 112 rotating display cubes; the show covered a wide range of topics that ranged from the creation of the world to the latest technological discoveries. In the Kino-automat theater, viewers were able to vote on how the film should proceed at five critical points, making it seem quite different on repeated viewings.

A wide range of Czechoslovakian artwork was on display. Much of it was religious in nature, which was somewhat surprising for a Communist country. One of the most spectacular was the *Trebechovice Bethleham*, an elaborate wooden Nativity scene carved by two artisans over 40 years that featured more than 2,000 figures, including 300 animated pieces. Young visitors enjoyed the Castle of Never-Never, a fanciful fairyland complete with castles, damsels in distress, and their rescuing heroes.

The newly independent nations of Guyana and Barbados joined together in a pavilion designed to not only tell of their histories but to also showcase their potential for modern businesses. The pavilion gained a degree of infamy when Millie, one of the macaws on display, revealed it had an X-rated vocabulary. It was equally fluent, and offensive, in both French and English. Millie was quickly banished but did star on postage stamps from Guyana for Christmas 1967.

Ceylon's pavilion was based on an audience hall in the city of Kandy, the last capital of the kings before they were deposed by Dutch and British invasions. On the outside, a ceramic mural celebrated the Kandy Esala Perahera, an annual religious festival. The displays inside included an intricate model of a tea plantation, examples of other products and tourism, and a café serving tea and staffed by hostesses in Ceylonese dress.

Eight large masts supported the steel netting and fabric roof of the Germany pavilion, a structure often referred to as a "space age circus tent." The unusual design allowed for a large amount of display space underneath the 100,000-square-foot canopy that did not require interior supports like conventional buildings. The pavilion represented West Germany; although there were numerous Communist pavilions at Expo 67, there was no mention of East Germany.

Unlike the other international pavilions, Germany did not have many displays on its history or tourism. Its focus was very much on the country's industrial might, with almost all the displays featuring new German products and designs. One notable exception was a replica of a 500-year-old printing press used by Johannes Gutenberg to print Bibles. It was demonstrated periodically by a staff member dressed in period garb.

Three giant brightly colored cubes marked the Venezuela pavilion. Painted in bright primary colors, the aluminum-sheathed cubes were a hit with both visitors and architecture critics. One cube held the usual display of history, tourism, and products; the second, a collection of Venezuelan art, live performances, and a restaurant; while the third contained a huge sculpture by Jesús Rafael Soto.

Ethiopia proved that a memorable pavilion could be built on a small budget. The structure was basically a 90-foot-tall tent surrounded be pillars styled after ancient tomb markers. At the top of the tent was a rotating golden lion, the symbol of Ethiopia, and nearby were two lion cubs, David and Bess, gifts from Emperor Haile Selassie. Complimentary Ethiopian coffee was a big hit, and a 26-minute film, *Man in Ethiopia*, was shown in the Queen of Sheba Theater.

A wide variety of items were imported from Australia for its pavilion. The outside was ringed by gardens of native plants and trees, and out back was an enclosure full of kangaroos and wallabies. There was even a display of more than 1,000 coral pieces from the Great Barrier Reef. The carpets were unbleached wool; a special machine vibrated visitors' shoes as they entered the building to keep the floors clean.

The most memorable part of the Australia pavilion was a group of 240 padded chairs with built-in speakers. Visitors could relax while listening to stories about Australia. There were 41 different stories spread across the chairs, and some guests stayed for quite a while trying to hear them all. It looks like one listener in this photograph found them a bit too relaxing. Color-coded cushions identified the English- and French-language chairs.

Fifteen newly independent African nations joined together in Africa Place, which was paid for by the Expo 67 corporation to help the nations participate. The pavilion roofs were all white when Expo opened but were painted during the season. Africa Place made headlines when a time bomb was discovered there on September 24, 1967, shortly before a scheduled visit by United Nations secretary general U Thant. It was defused just five minutes before it was set to detonate.

The lower level of the India pavilion was devoted to displays of life in India over the past 5,000 years. A circular theater used 1,200 slides on nine screens for *The Journey to India*, which focused more on the years since its independence in 1947. Upstairs, the focus shifted dramatically to modern India, with many examples of the products and technologies that were transforming the economy from an agricultural to an industrial basis.

Mexico was another country proud of its heritage and looking eagerly to the future. It featured a stone arch, said to be 1,000 years old, which once led to the Mayan city of Uxmal. At Expo, it was often used as the backdrop for a mariachi band. Inside the main building were exhibits of Mexican art and history that included a number of significant pre-Columbian pieces. The pavilion also housed a popular restaurant and bar.

Algeria, Egypt, and Kuwait shared a complex known as the Arab Group pavilion. When tensions flared due to the Six Day War between Egypt and Israel, Kuwait closed its pavilion on May 29, 1967, to protest the lack of support from Western nations for the Arabic side. It was the only country to withdraw from Expo during the six-month season.

Egypt stayed at Expo despite the conflict. The pavilion reflected the country's two names. The United Arab Republic was a short-lived union with Syria that had dissolved in 1961, but the name was carried over by Egypt until 1971. Exhibits ranged from replicas of items from the tomb of Tutankhamen to models of the new Aswan Dam. Numerous new Egyptian products were also on display.

Morocco used a colorful minaret to make its pavilion easier to find and to salute its Islamic heritage. Inside the main building, many of the displays stressed how Morocco had developed vast cities while the Western world was still in its early stages of development. Ancient artworks and scientific discoveries were shown, along with pictures of Morocco's natural beauty.

Intricately carved wooden doors gave a hint of what was inside the otherwise rather austere-looking Tunisia pavilion. A mosaic from a 2nd-century Roman villa was just one of the antiquities on display inside; other pieces were displayed alongside craftspeople who demonstrated traditional techniques used for rug weaving, tinwork, the building of elaborate birdhouses, and other goods. There were also performances by dancers, singers, and musicians in native costumes.

Thailand had two buildings for its exhibits. The one pictured on the left was a replica of an 18th-century Buddhist shrine and was moved to Expo 67 after being used at the 1964–1965 New York World's Fair. The second building, seen on the right, was a mixture of traditional and modern styles and featured displays of Thai crafts and products. A replica of a royal barge floated serenely in a pool next to an adjacent canal.

Burma used a similar approach for its pavilion. One building was styled after an ancient shrine for its main display area and was full of crafts and products, as well as displays proclaiming the country's progress in improving its agricultural industries and the lives of its citizens. The second building was a glass-walled restaurant that served a wide variety of Burmese dishes and drinks. Dance acts and musicians entertained between the two buildings.

The Sermons from Science pavilion was dedicated to proving the theory that scientific principles can be relied on as they are the work of a supreme creator. Films from the Moody Scientific Institute in Los Angeles were combined with live demonstrations, the most memorable of which was a man who allowed a million volts of electricity to pass through his body to ignite a torch held in his hand.

Six of Canada's leading chemical companies sponsored Kaleidoscope, which was dedicated to color in daily life. The 112 fins on the outside of the building created the illusion that the outer ring was rotating as visitors passed the structure. Inside, the show *Man and Color* used large mirrors to reflect colors and images just like a giant real-life kaleidoscope. A sophisticated sound system added to the experience.

The Judaism pavilion sprang to life after a group of rabbis was rebuffed in the effort to build a small synagogue at the Israel pavilion, as the Israeli government wanted to show it as a multi-faith nation. They responded by raising the money from Canada's Jewish community for a pavilion that traced many of the political and social battles the religion had faced, including the Holocaust and recent problems in the Middle East.

People who ignore "Don't Touch" signs were undoubtedly thrilled with the Canadian Pacific-Cominco pavilion. The two companies joined for an interactive exhibit area that encouraged visitors to explore all of their senses with displays that made noise, changed temperature, emitted odors, vibrated, and more. There was also a well-received film, *We Are Young*, a 20-minute adventure that used six screens as viewers followed the subjects on the road to maturity.

The Canadian Pulp and Paper pavilion was a whimsical, stylized forest of 44 evergreen trees, the tallest soaring to a height of 80 feet. Exhibits followed the development of paper, from its earliest days to the most recent huge mills that turned out this important product of Canadian industry. Guests could watch paper being made by hand and see examples of crafts created using paper products. They could also see a film on forest legends from around the world.

Next door was the Steel pavilion, which explored another important Canadian industry. Inside the 10-story building, guests were treated to the sights and smells of a steel mill, with examples of the raw products and the many steps needed to create steel. There was also a 350-seat theater with a film that showed how mankind has used natural resources and technology to reshape its world.

A large pavilion at more than seven acres, Man the Provider was usually the least crowded attraction. The exhibit warned of the need to increase farm production and efficiency to avoid famine. New births were occurring at a rate of two per second, swelling the population, but only three percent of the earth's surface is suitable for farming, making the need for agricultural improvements a necessity. On the lighter side, children were encouraged to interact with the animals on display.

Determined to make a good show at Expo 67 and to outdo the United States, the Soviet Union spent $15 million on its pavilion, second only to Canada. The sweeping roof was said to represent how Russia and the other Soviet states were taking off, both economically and in space. The building was a remarkable achievement, with the large glass sides affording tempting views of the exhibits inside, and some remarkable views of Île Notre-Dame from inside.

Long lines were very common at the USSR pavilion, which ranked as the most popular exhibit at Expo 67 with more than 15 million visitors passing through it. The building featured the largest restaurant at Expo, seating 1,100 guests, but many diners felt it was overpriced and it was the least successful part of the pavilion.

Most of the USSR displays were industrial in nature. In stark contrast to those in the United States pavilion, there was little mention made of the softer side of life. A few paintings were on display, and visiting folk groups entertained periodically, but the overall emphasis was very much on Soviet achievements in manufacturing and technology. The model seen here under the picture of Lenin showed some of the major utility systems across the country, such as power plants and gas lines.

The largest section of displays inside the USSR pavilion was a collection of spacecraft. Some of the displays were replicas, others were test vehicles, and some were models of proposed craft that never did fly. With the Space Race still in high gear, this was an especially popular part of the pavilion, and a simulated spaceflight attraction was almost always filled to capacity.

# *Four*

# LA RONDE

Although prior world's fairs and expositions had included their amusement zones as part of the main event, Expo 67 took a decidedly different approach. La Ronde was separated from the pavilions on Île Sainte-Hélène by the open expanses of Parc Hélène de Champlain. The area was accessible by tram, foot, Expo Express, hovercraft, helicopter, or private boat. This viewpoint was from the La Spirale observation tower.

Dolphin Lake made up about half of the 135-acre La Ronde site. This section of the island was created by the expansion of a tiny island previously named Île Ronde (Round Island), and the lake was formed when rocks were excavated from the area to connect it with Île Sainte-Hélène. The square area marked by floating logs was used for log-rolling demonstrations.

The lake was also used for water-skiing shows, boat shows, and canoe rides. The ski shows were organized by Marc Cloutier, one of Canada's leading skiers. A total of 748 performances were held, some including guest stars who were visiting Expo 67. One can only hope they were prepared for the undoubtedly chilly lake waters at the beginning and end of the Expo season.

Dolphin Lake was home to a full-sized replica of the *Grande Hermine*, the ship that carried Jacques Cartier to the area during his explorations in 1535. Several years after Expo ended, the ship was moved to the Cartier-Brébeuf National Historic Site in Quebec City. It later deteriorated and was destroyed. Pictured behind it is Le Village, a collection of shops, restaurants, and nightclubs in a re-creation of an old French Canadian village. The area was also used for dance and music performances by visiting groups.

Another re-creation of old Canada was Fort Edmonton-Pioneerland. Styled after the early days of Edmonton in Alberta, the fort promised visitors a taste of Canada's "Old West." The largest exhibit inside was the Golden Garter Saloon, complete with dancing girls. Smaller shops sold a wide variety of gifts, caricatures, and snacks. Men could also enjoy a haircut and a shave. Gunfights were held several times a day, much to the delight of the younger visitors.

The most popular attraction at Fort Edmonton was the log flume ride that jutted out into Dolphin Lake. The aquatic thrill ride was quite novel for 1967, as the first one anywhere had opened just four years earlier. The Expo edition was the first one in Canada. La Ronde is still open today, and the log flume ride is there as a tangible connection to the fun and excitement of Expo 67. The prices have gone up a bit though.

La Spirale is another surviving attraction from Expo 67 that is still in operation. Guests board the two-level observation car for a leisurely one-minute ride to the top of the 312-foot tower. The car rotates as it climbs and descends, which provides for wonderful views of the Expo site, the St. Lawrence River, and Montreal.

This view from La Spirale shows the Alcan Aquarium at the center and the maintenance yard for the Expo Express trains under the Jacques Cartier Bridge. The aquarium was intended as a permanent attraction and remained open when Expo 67 closed, but it ran into a series of problems, including the starvation deaths of several dolphins during a strike, and finally closed permanently in 1991.

Another view from La Spirale shows one of the many amusement park–type rides that were at La Ronde. Expo 67 publicity releases proclaimed that teams had traveled around the world to bring back the newest rides, and, indeed, many of them made their North American debuts at Expo. All of these rides, along with the games of chance in the colorful stands spread across the area, made La Ronde especially popular with teenage visitors.

The Sky Ride was another way to get some nice aerial views of the area. The ride also served a practical purpose as it was a convenient way to get across La Ronde, with one station at the main entrance and the other at Le Village on the far side of Dolphin Lake. Each of the colorful cars could carry up to four passengers on the 120-foot-high trip.

The 1,500-seat Garden of the Stars was used for a number of different purposes. During the day, it housed shows that varied from rock bands to poetry readings and table tennis exhibitions. In the afternoons, it converted to a teenage discotheque, and at night it featured a Las Vegas–style show. The theater even hosted a Grateful Dead concert on August 6, 1967. The building is still part of La Ronde today.

Gyrotron was a new $3 million ride designed especially for Expo 67 by noted British set designer Sean Kenny. Riders were treated to a simulated space voyage, then plunged into a pit of lava where a menacing monster awaited. The actual ride was less interesting than the description, though, and Gyrotron is best remembered for the 245-ton aluminum framework surrounding the pyramid-shaped building.

Nestled in front of the Gyrotron was the Youth Pavilion. Designed for audiences aged 16 to 30, it was sponsored by 33 youth organizations from across the country. There was a theater that showed student and experimental films, a game room, an open-air amphitheater, and, at night, a discotheque. Hundreds of different events were staged there, many featuring young artists and speakers from around the world.

*Laterna Magica* was a unique blend of film and live performers. Innovative techniques were used to make it seem as though some of the actors had left the film and come to life on the stage. After it became a hit when introduced at Expo 58 in Brussels, a second film was added for the Expo 67 experience. The show next appeared at HemisFair 68 in San Antonio, Texas, and is still being performed today in Prague.

Safari offered a very different entertainment experience than the rest of La Ronde. Guests journeyed into the "jungle" aboard a tram, on animal-drawn carts, or for even more fun, on a camel, an elephant, or even an ostrich. The riders were treated to views of a variety of animals, with the more dangerous ones kept at a safe distance. Piped-in jungle sounds made the adventure even more realistic.

Children's World featured rides for visitors ages four to nine. The young visitors could enjoy a number of scaled-down rides, including these antique cars and, in the background, *Old 99*, a cartoonish steam engine. There was a small roller coaster and a "ride in a tub" water attraction that is still in use today. The area also featured puppet shows, snack bars, and, not surprisingly, toy stores.

Adults probably enjoyed this antique carousel as much as the kids did. Named the Galloping, it was built in Belgium in 1885 and is said to be the oldest "galloping" carousel in the world. It came to Expo 67 after two seasons at the Belgium Village in the 1964–1965 New York World's Fair. After falling into disrepair, it was moved into storage for several years, but it was eventually restored and is part of La Ronde again today.

The carousel was part of the Carrefour International exhibit, a shopping and dining area with 40 shops and several well-received restaurants. Many of the venues were offshoots of the shops and restaurants in the main sections of Expo, giving visitors a chance to shop and dine in a less crowded environment. The Bavarian Beer Garden was especially popular and was one of the most successful restaurants at Expo. The area also included Expo's main post office.

La Ronde also featured a marina with moorings for 250 boats. Visitors could tie up there for up to eight days while they visited Expo. The building seen at the water's edge provided showers, laundry facilities, and supplies, and there was also a ship's chandler in case small repairs were needed. The marina was very popular and was often filled to capacity.

*Five*

# MAN AND HIS WORLD

Expo 67 was originally scheduled to close on October 27 to meet the BIE restriction of a six-month season. The date was a Friday, and the BIE later agreed to a two-day extension so the crowds could enjoy one last weekend. On October 7, though, Mayor Jean Drapeau made an announcement that was unprecedented in the history of world's fairs: Expo 67 would reopen the following year as a permanent attraction to be called Man and His World. Two earlier fairs had opened for a second season to recoup their expenses, but this was the first, and still only, time that an exposition would continue on indefinitely.

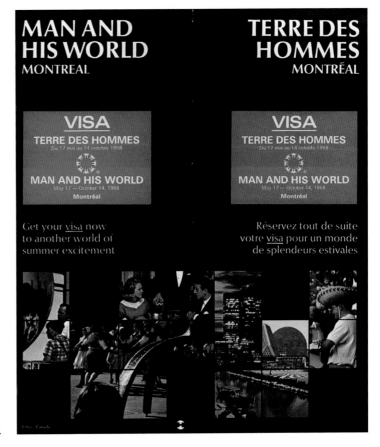

Drapeau used a bold strategy to create Man and His World by asking the various exhibitors to give their pavilions to the city. Most were more than happy to do so because it allowed them to avoid the cost of demolition. Some exhibitors decided to continue to operate their pavilions, but others simply turned over the keys and walked away. This made for some strange rebranding, such as when the Great Britain pavilion was reused in 1968 as the Cars of Yesteryear exhibit.

Part of the fun for each season of Man and His World was seeing how the pavilions had been transformed. This had been the Air Canada pavilion for Expo 67; in later years, it served as the Flight (1968), Communications (1969), Leonardo da Vinci (1970), Administration (1971), Flight (1972), Pakistan (1973–1976), and Vienna (seen here; 1977–1979) pavilions. A new guide map was an essential tool each year to reduce the confusion. A list of all the changes over the years is online at http://www.worldsfairphotos.com/expo67/man-and-his-world/index.htm.

Most of the Communist bloc nations decided to remove their pavilions, stating that they had previously planned to reduce their expenses by reusing or selling the buildings. The USSR brought its massive pavilion back to Moscow, where it is still used as an exhibition center, and the giant concrete slab where it once stood was used as a heliport for sightseeing tours. The area looked rather ragged in 1968 but was cleaned up for later years.

One of the strangest sights of Man and His World was this staircase left behind when the rest of the Czechoslovakia pavilion was dismantled. For some reason, it remained as a sort of observation tower that overlooked a mostly concrete area named the Terrace. The rest of the pavilion is now the Gordon Pinsent Centre for the Arts in Grand Falls, Newfoundland.

Some of the countries that had given up their pavilions decided to return in later seasons. The United States had deeded its giant geodesic dome to the City of Montreal, which operated it as an ecologically themed exhibit dubbed Biosphere, so when the US decided to return in 1970 it used the former Telephone Pavilion. In 1971, the United States returned to its original dome and then used the old Netherlands building for its final year in 1972.

With some exhibitors not staying on for Man and His World, new participants could become part of the show for far less than it would have cost them to build their own new pavilions. Ireland joined in 1968, for just one year, using the Economic Progress pavilion. The building had three new tenants in the following three years before it was closed permanently. Bulgaria joined in 1969 and moved through four different buildings over the next decade.

Some of the original designs made reusing the pavilions a bit of a challenge. What does one do, for example, with a pavilion shaped like a grove of trees? In the case of the Illusions pavilion, seen here in 1970, paint it in bright new colors and hope no one notices. The prior two years the building had been painted blue as the Police and Interpol pavilions.

By 1971, several of the structures were in poor shape as they had only been built to last for the six months of Expo 67 and not intended to survive the harsh Montreal winters. Some were closed that year for repairs, and the Mexico pavilion was demolished. A sign claimed the pavilion was "Closed this year to allow for an improved presentation next year," but it was not to be.

In 1972, Île Notre-Dame was closed when two labor strikes meant the pavilions could not be ready in time for the season. The deserted grounds made for an eerie sight as guests traveled across the island aboard the Minirail from Île Sainte-Hélène, which simply ran in a loup without stopping on Île Notre-Dame. There were hopes to reopen the area for 1973 but funds were diverted to the upcoming 1976 Olympics, and the area stayed shuttered.

Man and His World continued for several more years using the pavilions on Île Sainte-Hélène. A devastating fire started by a welder's torch ignited the plastic skin on the Biosphere (the former United States pavilion) on May 20, 1976. The building sat in ruins until it was repaired and reopened in 1995. The Biosphere continues to operate today and is a prized part of modern Montreal.

The former New York State pavilion saw new life in 1977 when it became a museum celebrating the 10th anniversary of Expo 67. Expo 67 + 10 described how Expo 67 had been born and built, using photographs and memorabilia that showcased some of the more popular pavilions. The exhibit also celebrated the growth of Montreal in the decade since Expo and included exhibits on future city projects.

In 1975, an exhibit showed how Île Notre-Dame would be reconfigured by building a canoeing and rowing basin for the 1976 Olympics. Plans to reopen some of the pavilions fell through when funding was unavailable due to Olympic cost overruns. The area was further altered in 1978 to add a Formula One racetrack. Several of the remaining pavilions made a final appearance in 1979 in an episode of television's *Battlestar Galactica* and in Robert Altman's film *Quintet*.

Visiting Man and His World in its later years was a bittersweet experience. It was possible to relive some of the fun of the Expo 67 days through the surviving pavilions, but the many closed-off areas were depressing. The once state-of-the-art Expo Express train system was shut down in 1973, and the cars were parked at the Île Notre-Dame station waiting for a new buyer. They were finally removed in 1979 and later cut up for scrap.

The year 1980 brought a new wave of activity to the old Expo site when it hosted Floralies, an international flower show, as part of that season's version of Man and His World. Several pavilions were reopened and others demolished to provide more open space. Crowds returned to the previously deserted expanse of Île Notre-Dame to tour gardens that were comprised of flowers and shrubs from around the world. Seen here is the former Asbestos Plaza, now awash in a sea of color.

Man and His World limped through to its final season in 1984. By that time, the attendance figures had declined to the point that it was no longer feasible to continue. Some of the pavilions were to be retained for use in a science museum, but when those plans fell through, the site was cleared in 1986 and 1987. When the demolition work was over, the site was reopened as a city park, now named Parc Jean-Drapeau in honor of the visionary mayor who had brought Expo to life. Some of the Expo buildings have survived, including the France and Quebec pavilions, which are now the Montreal Casino, and the Jamaica pavilion, now available as a meeting hall. Calder's massive sculpture *Man* is still there but on a new site. The Biosphere and La Ronde continue to operate, keeping the Expo experience alive. Modern world's fairs have become disposable events, leaving very little behind to mark their passage; it is very unlikely there will ever be an exposition with the staying power and legacy of Expo 67. (Courtesy of Roger La Roche.)

Visit us at
www.arcadiapublishing.com

Visitez nous à
www.arcadiapublishing.com